Asteroid Of Fear

by

RAYMOND Z. GALLUN

Asteroid Of Fear
by Raymond Z. Gallun

Copyright © 2023

ISBN: 978-93-59324-79-1

Published by

DOUBLE 9 BOOKS

2/13-B, Ansari Road, Daryaganj
New Delhi – 110002
info@double9books.com
www.double9books.com
Tel. 011-40042856

ABOUT THE AUTHOR

Raymond Zinke Gallun (March 22, 1911 – April 2, 1994) was a science fiction author from the United States. Gallun (rhymes with "balloon") was the son of Adolph and Martha Zinke Gallun and was born in Beaver Dam, Wisconsin. In 1928, he graduated from high school. He dropped out of college after a year and traveled through Europe, living as a vagabond and doing a variety of jobs all over the world in the years leading up to World War II. At the age of 16, Gallun penned his first two stories, "The Space Dwellers" and "The Crystal Ray" (both published in 1929). He was a member of the early sci-fi pulp writers that popularized the genre. In the 1930s, he sold numerous popular stories to pulp publications. His first notable story was "Old Faithful" (1934). "The Gentle Brain" appeared in "Science Fiction Quarterly" under the pen name Arthur Allport. People Minus X, his debut book, was released by Simon & Schuster in 1957, followed by The Planet Strappers (Pyramid) in 1961. The Best of Raymond Z. Gallun, published by Ballantine in 1978, is a selection of his early work. Gallun received the I-CON Lifetime Achievement Award at I-CON IV in 1985; the award was eventually renamed The Raymond Z. Gallun Award.

All space was electrified as that harsh challenge rang out ... but John Endlich hesitated. For he saw beyond his own murder — saw the horror and destruction his death would unleash — and knew he dared not fight back!

The space ship landed briefly, and John Endlich lifted the huge Asteroids Homesteaders Office box, which contained everything from a prefabricated house to toothbrushes for his family, down from the hold-port without help or visible effort.

In the tiny gravity of the asteroid, Vesta, doing this was no trouble at all. But beyond this point the situation was — bitter.

His two kids, Bubs, seven, and Evelyn, nine — clad in space-suits that were slightly oversize to allow for the growth of young bodies — were both bawling. He could hear them through his oxygen-helmet radiophones.

Around him, under the airless sky of space, stretched desolation that he'd of course known about beforehand — but which now had assumed that special and terrible starkness of reality.

At his elbow, his wife, Rose, her heart-shaped face and grey eyes framed by the wide face-window of her armor, was trying desperately to choke back tears, and be brave.

"Remember — we've *got* to make good here, Johnny," she was saying. "Remember what the Homesteaders Office people told us — that with modern equipment and the right frame of mind, life can be nice out here. It's worked on other asteroids. What if we are the first farmers to come to Vesta?... Don't listen to those crazy miners! They're just kidding us! Don't listen to them! And don't, for gosh sakes, get sore...."

Rose's words were now like dim echoes of his conscience, and of his recent grim determination to master his hot temper, his sensitiveness, his wanderlust, and his penchant for poker and the social glass — qualities of an otherwise agreeable and industrious nature, that, on Earth, had always been his undoing. Recently, back in Illinois, he had even spent six months in jail for all but inflicting murder with his bare fists on a bullying neighbor whom he had caught whipping a horse. Sure — but during those six months his farm, the fifth he'd tried to run in scattered parts of North America, had gone to weeds in spite of Rose's valiant efforts to take care of it alone....

Oh, yes — the lessons of all that past personal history should be strong in his mind. But now will power and Rose's frightened tones of wisdom both seemed to fade away in his brain, as jeering words from another source continued to drive jagged splinters into the weakest portion of his soul:

"Hi, you hydroponic pun'kin-head!... How yuh like your new claim?... Nice, ain't it? How about some fresh turnips?... Good luck, yuh greenhorn.... Hiyuh, papa! Tied to baby's diaper suspenders!... Let the poor dope alone, guys.... Snooty.... Won't take our likker, hunh? Won't take our money.... Wifey's boy! Let's make him sociable.... Haw-Haw-haw.... Hydroponic pun'kin-head!..."

It was a medley of coarse voices and laughter, matching the row of a dozen coarse faces and grins that lined the view-ports of the ship. These men were asteroid miners, space-hardened and space-twisted. They'd been back to Earth for a while, to raise hell and freshen up, and spend the money in their then-bulging pockets. Coming out again from Earth, across the orbit of Mars to the asteroid belt, they had had the Endlichs as fellow passengers.

John Endlich had battled valiantly with his feebler side, and with his social inclinations, all through that long, dreary voyage, to keep clear of the inevitable griefs that were sure to come to a chap like himself from involvement with such characters. In the main, it had been a rather tattered victory. But now, at the final moment of bleak anticlimax, they took their revenge in guffaws and ridicule, hurling the noise at him through the radiophones of the space-suit helmets that they held in their laps — space-suits being always kept handy beneath the traveler-seats of every interplanetary vessel.

"... Haw-haw-haw! Drop over to our camp sometime for a little drink, and a little game, eh, pantywaist? Tain't far. Sure — just drop in on us when the pressure of domesticity in this beootiful country gets you down.... When the turnips get you down! Haw-haw-haw! Bring the wife along.... She's kinda pretty. Ought to have a man-size fella.... Just ask for me — Alf Neely! Haw-haw-haw!"

Yeah, Alf Neely was the loudest and the ugliest of John Endlich's baiters. He had gigantic arms and shoulders, small squinty eyes, and a pendulous nose. "Haw-haw-haw!..."

And the others, yelling and hooting, made it a pack: "Man — don't he wish he was back in Podunk!... What! — no tomatas, Dutch?... What did they tell yuh back at the Homestead office in Chicago? — that we were in de-e-esperate need of fresh vegetables out here? Well, where are they, papa?... Haw-haw-haw!..."

Under the barrage John Endlich's last shreds of common-sense were all but blotted out by the red murk of fury. He was small and broad — a stolid-looking thirty-two years old. But now his round and usually placid face was as red as a fiery moon, and his underlip curled in a snarl. He might have taken the savage ribbing more calmly. But there was too much grim fact behind what these asteroid miners said. Besides, out here he had thought that he would have a better chance to lick the weaknesses in himself—because he'd *have* to work to keep his family alive; because he'd been told that there'd be no one around to distract him from duty. Yah! The irony of that, now, was maddening.

For the moment John Endlich was speechless and strangled—but like an ignited firecracker. Uhunh—ready to explode. His hard body hunched, as if ready to spring. And the baiting waxed louder. It was like the yammering of crows, or the roar of a wild surf in his ears. Then came the last straw. The kids had kept on bawling—more and more violently. But now they got down to verbal explanations of what they thought was the matter:

"Wa-aa-aa-a-ahh-h! Papa—we wanna-go-o-o—hom-m-mm-e!..."

The timing could not have been better—or worse. The shrieks and howls of mirth from the miners, a moment ago, were as nothing to what they were now.

"Ho-ho-ho! Tell it to Daddy, kids!... Ho-ho-ho! That was a mouthful.... Ho-ho-ho-ho! Wow!..."

There is a point at which an extremity of masculine embarrassment can lead to but one thing—mayhem. Whether the latter is to be inflicted on the attacked or the attacker remains the only question mark.

"I'll get you, Alf Neely!" Endlich snarled. "Right now! And I'll get all the damned, hell-bitten rest of you guys!"

Endlich was hardly lacking in vigor, himself. Like a squat but streamlined fighting rooster, rendered a hundred times more agile by the puny gravity, he would have reached the hold-port threshold in a single lithe skip—had not Rose, despairing, grabbed him around the middle to restrain him. Together they slid several yards across the dried-out surface of the asteroid.

"Don't, Johnny—please don't!" she wailed.

Her begging could not have stopped him. Nor could her physical interference—for more than an instant. Nor could his conscience, nor his recent determination to keep out of trouble. Not the certainty of being torn limb from limb, and not hell, itself, could have held him back, anymore, then.

Yet he was brought to a halt. It certainly wasn't cowardice that accomplished this. No.

Suddenly there was no laughter among the miners. But in a body they arose from their traveler-seats aboard the ship. Suddenly there was no more humor in their faces beyond the view-ports. They were itching to be assaulted. The glitter in Alf Neely's small eyes was about as reassuring as the glitter in the eyes of a slightly prankish gorilla.

"We're waitin' for yuh, Mr. Civilization," he rumbled softly.

After that, all space was still — electrified. The icy stars gleamed in the black sky. The shrunken sun looked on. And John Endlich saw beyond his own murder. To the thought of his kids — and his wife — left alone out here, hundreds of millions of miles from Earth, and real law and order — with these lugs. These guys who had been starved emotionally, and warped inside by raw space. Coldness crawled into John Endlich's guts, and seemed to twist steel hooks there, making him sick. The silence of a vacuum, and of unthinkable distances, and of ghostly remains which must be left on this fragment of a world that had blown up, maybe fifty million or more years ago, added its weight to John Endlich's feelings.

And for his family, he was scared. What hell could not have accomplished, became fact. His almost suicidal impulse to inflict violence on his tormenters was strangled, bottled-up — brutally repressed, and left to impose the pangs of neurosis on his tormented soul. Narrowing domesticity had won a battle.

Except, of course, that what he had already said to Alf Neely and Friends was sufficient to start the Juggernaut that they represented, rolling. As he picked himself and Rose up from the ground, he saw that the miners were grimly donning their space-suits, in preparation to their coming out of the ship to lay him low.

"Oh — tired, hunh, Pun'kin-head?" Alf Neely growled. "It don't matter, Dutch. We'll finish you off without you liftin' a finger!"

In John Endlich the rage of intolerable insults still seethed. But there was no question, now, of outcome between it and the brassy taste of danger on his tongue. He knew that even knuckling down, and changing from man to worm to take back his fighting words, couldn't do any good. He felt like a martyr, left with his family in a Roman arena, while the lions approached. His butchery was as good as over....

Reprieve came presumably by way of the good-sense of the pilot of the space ship. The hold-port was closed abruptly by a mechanism that could be operated only from the main control-board. The rocket jets of the craft emitted a single weak burst of flame. Like a boulder grown agile and flighty, the ship leaped from the landscape, and arced outward toward the stars, to curve around the asteroid and disappear behind the scene's jagged brim.

The craft had gone to make its next and final stop—among the air-domes of the huge mining camp on the other side of Vesta—the side of torn rocks and rich radioactive ores.

But before the ship had vanished from sight, John Endlich heard Alf Neely's grim promise in his helmet radiophones: "We'll be back tonight, Greenhorn. Lots of times we work night-shift—when it's daytime on this side of Vesta. We'll be free. Stick around. I'll rub what's left of you in the dust of your claim!"

Endlich was alone, then, with the fright in his wife's eyes, the squalling of his children, and his own abysmal disgust and worry.

For once he ceased to be a gentle parent. "Bubs! Evelyn!" he snapped. "Shud-d-d—up-p-p!..."

The startled silence which ensued was his first personal victory on Vesta. But the silence, itself, was an insidious enemy. It made his ears ring. It made even his audible pulsebeats seemed to ache. It bored into his nerves like a drill. When, after a moment, Rose spoke quaveringly, he was almost grateful:

"What do we do, Johnny? We've still got to do what we're supposed to do, don't we?"

Whereupon John Endlich allowed himself the luxury and the slight relief of a torrent of silent cussing inside his head. Damn the obvious questions of women! Damn the miners. Damn the A.H.O.—the Asteroids Homesteaders Office—and their corny slogans and posters, meant to hook suckers like himself! Damn his own dumb hide! Damn the mighty urge to get drunk! Damn all the bitter circumstances that made doing so impossible. Damn! Damn! Damn!

Finished with this orgy, he said meekly: "I guess so, Hon."

All members of the Endlich family had been looking around them at the weird Vestal landscape. Through John Endlich's mind again there flashed a picture of what this asteroid was like. At the Asteroids Homesteaders' School in Chicago, where his dependents and he had been given several weeks of orientation instruction, suitable to their separate needs, he had been shown diagrams and photographs of Vesta. Later, he had of course seen it from space.

It was not round, like a major planet or most moons. Rather, it was like a bomb-fragment; or even more like a shard of a gigantic broken vase. It was several hundred miles long, and half as thick. One side of it—this side—was curved; for it had been a segment of the surface of the shattered planet from which all of the asteroids had come. The other side was jagged and broken, for it had been torn from the mesoderm of that tortured mother world.

From the desolation of his own thoughts, in which the ogre-form of Alf Neely lurked with its pendent promise of catastrophe soon to come, and from his own view of other desolation all around him, John Endlich was suddenly distracted by the comments of his kids. All at once, conforming to the changeable weather of children's natures regardless of circumstance, their mood had once more turned bright and adventurous.

"Look, Pop," Bubs chirped, his round red face beaming now from his helmet face-window, in spite of his undried tears. "This land all around here was fields once! You can even see the rows of some kind of stubble! Like corn-stubble! And over there's a—a—almost like a fence! An' up there is hills with trees on 'em—some of 'em not even knocked over. But everything is all dried-out and black and grey and dead! Gosh!"

"We can see all that, Dopey!" Evelyn, who was older, snapped at Bubs. "We know that something like people lived on a regular planet here, awful long ago. Why don't you look over the other way? There's the house—and maybe the barn and the sheds and the old garden!"

Bubs turned around. His eyes got very big. "Oh! O-ooh-h-h!" he gasped in wonder. "Pop! Mom! Look! Don't you see?..."

"Yeah, we see, Bubs," John Endlich answered.

For a long moment he'd been staring at those blocklike structures. One—maybe the house—was of grey stone. It had odd, triangular windows, which may once have been glazed. Some of the others were of a blackened material—perhaps cellulose. Wood, that is. All of the buildings were pushed askew, and partly crumpled from top to bottom, like great cardboard cartons that had been half crushed.

Endlich's imagination seemed forced to follow a groove, trying to picture that last terrible moment, fifty-million years ago. Had the blast been caused by natural atomic forces at the heart of the planet, as one theory claimed? Or had a great bomb, as large as an oversized meteor, come self-propelled from space, to bury itself deep in that ancient world? A world as big as Mars, its possible enemy—whose weird inhabitants had been wiped out, in a less spectacular way, perhaps in the same conflict?

Endlich's mind grabbed at that brief instant of explosion. The awful jolt, which must have ended all consciousness, and all capacity for eyes to see what followed. Perhaps there was a short and terrible passing of flame. But in swift seconds, great chunks of the planet's crust must have been hurled outward. In a moment the flame must have died, dissipated with the suddenly vanishing atmosphere, into the cold vacuum of the void. Almost instantly, the sky, which had been deep blue before, must have turned to its present black, with the voidal stars blazing. There had been no air left to sustain combustion, so buildings and trees had not continued to burn, if there had been time at all to ignite them. And, with the same swiftness, all remaining artifacts and surface features of this chip of a world's crust that was Vesta, had been plunged into the dual preservatives of the interplanetary regions — deep-freeze and all but absolute dryness. Yes — the motion of the few scattered molecules in space was very fast — indicating a high temperature. But without substance to be hot, there can be no heat. And so few molecules were there in the void, that while the concept of a "hot" space remained true, it became tangled at once with the fact that a *practically* complete vacuum can have *practically* no temperature. Which meant — again in practice — all but absolute zero.

John Endlich knew. He'd heard the lectures at the Homesteaders' School. Here was a ghost-land, hundreds of square miles in extent — a region that had been shifted in a few seconds, from the full prime of life and motion, to moveless and timeless silence. It was like the mummy of a man. In its presence there was a chill, a revulsion, and yet a fascination.

The kids continued to jabber — more excitedly now than before. "Pop! Mom!" Bubs urged. "Let's go look inside them buildings! Maybe the *things* are still there! The people, I mean. All black and dried up, like the one in the showcase at school; four tentacles they had instead of arms and legs, the teacher said!"

"Sure! Let's go!" Evelyn joined in. "I'm not scared to!"

Yeah, kids' tastes could be pretty gruesome. When you thought most that you had to shelter them from horror, they were less bothered by it than you were. John Endlich's lips made a sour line.

"Stay here, the pair of you!" Rose ordered.

"Aw — Mom —" Evelyn began to protest.

"You heard me the first time," their mother answered.

John Endlich moved to the great box, which had come with them from Earth. The nervous tension that tore at him — unpleasant and chilling, driving him toward straining effort — was more than the result of the shameful and embarrassing memory of his very recent trouble with Alf Neely and Companions, and the certainty of more trouble to come from that source. For there was another and even worse enemy. Endlich knew what it was —

The awful silence.

He still looked shamefaced and furious; but now he felt a gentler sharing of circumstances. "We'll let the snooping go till later, kids," he growled. "Right now we gotta do what we gotta do —"

The youngsters seemed to join up with his mood. As he tore the pinchbar, which had been conveniently attached to the side of the box, free of its staples, and proceeded to break out supplies, their whimsical musings fell close to what he was thinking.

"Vesta," Evelyn said. "They told us at school—remember? Vesta was the old Roman goddess of hearth and home. Funny—hunh—Dad?"

Bubs' fancy was vivid, too. "Look, Pop!" he said again, pointing to a ribbon of what might be concrete, cracked and crumpled as by a terrific quake, curving away toward the hills, and the broken mountains beyond. "That was a road! Can't you almost hear some kinda cars and trucks goin' by?"

John Endlich's wife, helping him open the great box, also had things to say, in spite of the worry showing in her face. She touched the dessicated soil with a gauntleted hand. "Johnny," she remarked wonderingly. "You can see the splash-marks of the last rain that ever fell here—"

"Yeah," Endlich growled without any further comment. Inside himself, he was fighting the battle of lost things. The blue sky. The shifting beauty of clouds in sunshine. The warm whisper of wind in trees. The rattle of traffic. The babble of water. The buzz of insects. The smell of flowers. The sight of grass waving.... In short, all the evidences of life.

"A lot of things that was here once, we'll bring back, won't we, Pop?" Bubs questioned with astonishing maturity.

"Hope so," John Endlich answered, keeping his doubts hidden behind gruffness. Maybe it was a grim joke that here and now every force in himself was concentrated on substantial objectives—to the exclusion of his defects. The drive in him was to end the maddening silence, and to rub out the mood of harsh barrenness, and his own aching homesickness, by struggling to bring back a little beauty of scenery, and a little of living motion. It was a civilized urge, a home-building urge, maybe a narrow urge. But how could anybody stand being here very long, unless such things were done? If they ever could be. Maybe, willfully, he had led himself into a grimmer trap than it had even seemed to be—or than he had ever wanted....

Inside his space suit, he had begun to sweat furiously. And it was more because of the tension of his nerves than because of the vigor with which he plied his pinchbar, doing the first task which had to be done. Steel ribbons were snapped, nails were yanked silently from the great box, boards were jerked loose.

In another minute John Endlich and his wife were setting up an airtight tent, which, when the time came, could be inflated from compressed-air bottles. They worked somewhat awkwardly, for their instruction period had been brief, and they were green; but the job was speedily finished. The first requirement—shelter—was assured.

Digging again into the vast and varied contents of the box, John Endlich found some things he had not expected—a fine rifle, a pistol and ammunition. At which moment an ironic imp seemed to sit on his shoulder, and laugh derisively. Umhm-m—the Asteroids Homesteaders Office had filled these boxes according to a precise survey of the needs of a peaceful settler on Vesta.

It was like Bubs, with the inquisitiveness of a seven-year-old, to ask: "What did they think we needed guns for, when they knew there was no rabbits to shoot at?"

"I guess they kind of suspected there'd be guys like Alf Neely, son," John Endlich answered dryly. "Even if they didn't tell us about it."

The next task prescribed by the Homesteaders' School was to secure a supply of air and water in quantity. Again, following the instructions they had received, the Endlichs uncrated and set up an atom-driven drill. In an hour it had bored to a depth of five-hundred feet. Hauling up the drill, Endlich lowered an electric heating unit on a cable from an atomic power-cell, and then capped the casing pipe.

Yes, strangely enough there was still sufficient water beneath the surface of Vesta. Its parent planet, like the Earth, had had water in its crust, that could be tapped by means of wells. And so suddenly had Vesta been chilled in the cold of space at the time of the parent body's explosion, that this water had not had a chance to dissipate itself as vapor into the void, but had been frozen solid. The drying soil above it had formed a tough shell, which had protected the ice beneath from disappearance through sublimation...

Drill down to it, melt it with heat, and it was water again, ready to be pumped and put to use.

And water, by electrolysis, was also an easy source of oxygen to breathe.... The soil, once thawed over a few acres, would also yield considerable nitrogen and carbon dioxide—the makings of many cubic meters of atmosphere. The A.H.O. survey expeditions, here on Vesta and on other similar asteroids which were crustal chips of the original planet, had done their work well, pathfinding a means of survival here.

When John Endlich pumped the first turbid liquid, which immediately froze again in the surface cold, he might, under other, better circumstances, have felt like cheering. His well was a success. But his tense mind was racing far ahead to all the endless tasks that were yet to be done, to make any sense at all out of his claim. Besides, the short day—eighteen hours long instead of twenty-four, and already far advanced at the time of his tumultuous landing—was drawing to a close.

"It'll be dark here mighty quick, Johnny," Rose said. She was looking scared, again.

John Endlich considered setting up floodlights, and working on through the hours of darkness. But such lights would be a dangerous beacon for prowlers; and when you were inside their area of illumination, it was difficult to see into the gloom beyond.

Still, one did not know if the mask of darkness did not afford a greater invitation to those with evil intent. For a long moment, Endlich was in an agony of indecision. Then he said:

"We'll knock off from work now—get in the tent, eat supper, maybe sleep..."

But he was remembering Neely's promise to return tonight.

In another minute the small but dazzling sun had disappeared behind the broken mountains, as Vesta, unspherical and malformed, tumbled rather than rotated on its center of gravity. And several hours later, amid heavy cooking odors inside the now inflated plastic bubble that was the tent, Endlich was sprawled on his stomach, unable, through well-founded worry, even to remove his space suit or to allow his family to do so, though there was breathable air around them. They lay with their helmet face-windows open. Rose and Evelyn breathed evenly in peaceful sleep.

Bubs, trying to be very much a man, battled slumber and yawns, and kept his dad company with scraps of conversation. "Let 'em come, Pop," he said cheerfully. "Hope they do. We'll shoot 'em all. Won't we, pop? You got the rifle and the pistol ready, Pop...."

Yes, John Endlich had his guns ready beside him, all right—for what it was worth. He wished wryly that things could be as simple as his hero-worshipping son seemed to think. Thank the Lord that Bubs was so trusting, for his own peace of mind—the prankish and savage nature of certain kinds of men, with liquor in their bellies, being what it was. For John Endlich, having been, on occasion, mildly kindred to such men, was well able to understand that nature. And understanding, now, chilled his blood.

Peering from the small plastic windows of the tent, he kept watching for hulking black shapes to silhouette themselves against the stars. And he listened on his helmet phones, for scraps of telltale conversation, exchanged by short-range radio by men in space armor. Once, he thought he heard a grunt, or a malicious chuckle. But it may have been just vagrant static.

Otherwise, from all around, the stillness of the vacuum was absolute. It was unnerving. On this airless piece of a planet, an enemy could sneak up on you, almost without stealth.

Against that maddening silence, however, Bubs presently had a helpful and unprompted suggestion: "Hey, Pop!" he whispered hoarsely. "Put the side of your helmet against the tent-floor, and listen!"

John Endlich obeyed his kid. In a second cold sweat began to break out on his body, as intermittent thudding noises reached his ear. In the absence of an atmosphere, sounds could still be transmitted through the solid substance of the asteroid.

It took Endlich a moment to realize that the noises came, not from nearby, but from far away, on the other side of Vesta. The thudding was vibrated straight through many miles of solid rock.

"It's nothing, Bubs," he growled. "Nothing but the blasting in the mines."

Bubs said "Oh," as if disappointed. Not long thereafter he was asleep, leaving his harrassed sire to endure the vigil alone. Endlich dared not doze off, to rest a little, even for a moment. He could only wait. If an evil visitation came — as he had been all but sure it must — that would be bad, indeed. If it didn't come — well — that still meant a sleepless night, and the postponement of the inevitable. He couldn't win.

Thus the hours slipped away, until the luminous dial of the clock in the tent — it had been synchronized to Vestal time — told him that dawn was near. That was when, through the ground, he heard the faint scraping. A rustle. It might have been made by heavy space-boots. It came, and then it stopped. It came again, and stopped once more. As if skulking forms paused to find their way.

Out where the ancient and ghostly buildings were, he saw a star wink out briefly, as if a shape blocked the path of its light. Then it burned peacefully again. John Endlich's hackles rose. His fists tightened on both his rifle and pistol.

He fixed his gaze on the great box, looming blackly, the box that contained the means of survival for his family and himself, as if he foresaw the future, a moment away. For suddenly, huge as it was, the box rocked, and began to move off, as if it had sprouted legs and come alive.

John Endlich scrambled to action. He slammed and sealed the face-windows of the helmets of the members of his family, to protect them from suffocation. He did the same for himself, and then unzipped the tent-flap. He darted out with the outrushing air.

This was a moment with murder poised in every tattered fragment of it. John Endlich knew. Murder was engrained in his own taut-drawn nerves, that raged to destroy the trespassers whose pranks had passed the level of practical humor, and become, by the tampering with vital necessities, an attack on life itself. But there was a more immediate menace in these space-twisted roughnecks.... Strike back at them, even in self-defense, and have it proven!

He had not the faintest doubt who they were — even though he could not see their faces in the blackness. Maybe he should lay low — let them have their way.... But how could he — even apart from his raging temper, and his honor as a man — when they were making off with his family's and his own means of survival?

He had to throw Rose and the kids into the balance — risking them to the danger that he knew lay beyond his own possible ignoble demise. He did just that when he raised his pistol, struggling against the awful impulse of the rage in him — lifted it high enough so that the explosive bullets that spewed from it would be sure to pass over the heads of the dark silhouettes that were moving about.

"Damn you, Neely!" Endlich yelled into his helmet mike, his finger tightening on the trigger. "Drop that stuff!"

At that moment the sun's rim appeared at the landscape's jagged edge, and on this side of airless Vesta complete night was transformed to complete day, as abruptly as if a switch had been turned.

Alf Neely and John Endlich blinked at each other. Maybe Neely was embarrassed a little by his sudden exposure; but if he was, it didn't show. Probably the bully in him was scared; but this he covered in a common manner — with a studiedly easy swagger, and a bravado that was not good sense, but bordered on childish recklessness. Yet he had a trump card — by the aggressive glint in his eyes, and his unpleasant grin, Endlich knew that Neely knew that he was afraid for his wife, and wouldn't start anything unless driven and goaded sheerly wild. Even now, they were seven to his one.

"Why, good morning, Neighbor Pun'kin-head!" Neely crooned, his voice a burlesque of sweetness. "Glad to oblige!"

He hurled the great box down. As he did so, something glinted in his gloved paw. He flicked it expertly into the open side of the wooden case which contained so many things that were vital to the Endlichs —

It was only a tiny nuclear priming-cap, and the blast was feeble. Even so, the box burst apart. Splintered crates, sealed cans, great torn bundles and what not, went skittering far across the plain in every direction, or were hurled high toward the stars, to begin falling at last with the laziness of a descending feather.

Neely and his companions hadn't attempted to move out of the way of the explosion. They only rolled with its force, protected by their space suits. Endlich rolled, too, helplessly, clutching his pistol and rifle: still, by some superhuman effort, he managed to regain his feet before the far more practiced Neely, who was hampered, no doubt, by a few too many drinks, had even stopped rolling. But when Neely got up, he had drawn his blaster, a useful tool of his trade, but a hellish weapon, too, at short range.

Still, Endlich retained the drop on him.

Alf Neely chuckled. "Fourth of July! Hallowe'en, Dutch," he said sweetly. "What's the matter? Don't you think it's fun? Honest to gosh — you just ain't neighborly!"

Then he switched his tone. It became a soft snarl that didn't alter his insolent and confident smirk — and a challenge. He laughed derisively, almost softly. "I dare you to try to shoot straight, pal," he said. "Even you got more sense than that."

And John Endlich was spang against his terrible, blank wall again. Seven to one. Suppose he got three. There'd be four left — and more in the camp. But the four would survive him. Space crazy lugs. Anyway half drunk. Ready to hoot at the stars, even, if they found no better diversion. Ready to push even any of their own bunch around who seemed weaker than they. For spite, maybe. Or just for the lid-blowing hell of it — as a reaction against the awful confinement of being out here.

"I was gonna smear you all over the place, Greenhorn," Neely rumbled. "But maybe this way is more fun, hunh? Maybe we'll be back tonight. But don't wait up for us. Our best regards to your sweet — family."

John Endlich's blazing and just rage was strangled by that same crawling dread as before, as he saw them arc upward and away, propelled by the miniature drive-jets attached to the belts of their space-suits. Their return to camp, hundreds of miles distant, could be accomplished in a couple of minutes.

Rose and the kids were crouched in the deflated tent. But returning there, John Endlich hardly saw them. He hardly heard their frightened questions.

To the trouble with Neely, he could see no end — just one destructive visitation following another. Maybe, already, mortal damage had been done. But Endlich couldn't lie down and quit, any more than a snake, tossed into a fire, could stop trying to crawl out of it, as long as life lasted. Whether doing so made sense or not, didn't matter. In Endlich was the savage energy of despair. He was fighting not just Neely and his crowd, but that other enemy — which was perhaps Neely's main trouble, too. Yeah — the stillness, the nostalgia, the harshness.

"No — don't want any breakfast," he replied sharply to Rose' last question. "Gotta work...."

He was like an ant-swarm, rebuilding a trampled nest — oblivious to the certainty of its being trampled again. First he scrambled and leaped around, collecting his scattered and damaged gear. He found that his main atomic battery — so necessary to all that he had to do — was damaged and unworkable. And he had no hope that he could repair it. But this didn't stop his feverish activity.

Now he started unrolling great bolts of a transparent, wire-strengthened plastic. Patching with an adhesive where explosion-rents had to be repaired, he cut hundred-yard strips, and, with Rose's help, laid them edge to edge and fastened them together to make a continuous sheet. Next, all around its perimeter, he dug a shallow trench. The edges of the plastic were then attached to massive metal rails, which he buried in the trench.

"Sealed to the ground along all the sides, Honey," he growled to Rose. "Next we fit in the airlock cabinet, at one corner. Then we've got to see if we can get up enough air to inflate the whole business. That's the tough part — the way things are...."

By then the sun was already high. And Endlich was panting raggedly — mostly from worry. After the massive airlock was in place, they attached their electrolysis apparatus to the small atomic battery, which had been used to run the well-driller. The well was in the area covered by the sheet of plastic, which was now propped up here and there with long pieces of board from the great box. Over their heads, the tough, clear material sagged like a tent-roof which has not yet been run up all the way on its poles.

Sluggishly the electrolysis apparatus broke down the water, discharging the hydrogen as waste through a pipe, out over the airless surface of Vesta — but freeing the oxygen under the plastic roof. Yet from the start it was obvious that, with insufficient electric power, the process was too slow.

"And we need to use heat-coils to thaw the ground, Johnny," Rose said. "And to keep the place warm. And to bring nitrogen gas up out of the soil. The few cylinders of the compressed stuff that we've got won't be enough to make a start. And the carbon dioxide...."

So John Endlich had to try to repair that main battery. He thought, after a while, that he might succeed — in time. But then Rose opened the airlock, and the kids came in to bother him. With all the triumph of a favorite puppy dragging an over-ripe bone into the house, Bubs bore a crooked piece of a black substance, hard as wood and more gruesome than a dried and moldy monkey-pelt.

"A tentacle!" Evelyn shrilled. "We were up to those old buildings! We found the people! What's left of them! And lots of stuff. We saw one of their cars! And there was lots more. Dad—you gotta come and see!..."

Harassed as he was, John Endlich yielded—because he had a hunch, an idea of a possibility. So he went with his children. He passed through a garden, where a pool had been, and where the blackened remains of plants still projected from beds of dried soil set in odd stone-work. He passed into chambers far too low for comfortable human habitation. And what did he know of the uses of most of what he saw there? The niches in the stone walls? The slanting, ramplike object of blackened wood, beside which three weird corpses lay? The glazed plaque on the wall, which could have been a religious emblem, a calendar of some kind, a decoration, or something beyond human imagining? Yeah—leave such stuff for Cousin Ernest, the school teacher—if he ever got here.

In the cylindrical stone shed nearby, John Endlich had a look at the car—low slung, three-wheeled, a tiller, no seats. Just a flat platform. All he could figure out about the motor was that steam seemed the link between atomic energy and mechanical motion.

Beyond the car was what might be a small tractor. And a lot of odd tools. But the thing which interested him most was the pattern of copper ribbons, insulated with a heavy glaze, similar to that which he had seen traversing walls and ceiling in the first building he had entered. Here, as before, they connected with queer apparatus which might be stoves and non-rotary motors, for all he knew. And also with the globes overhead.

The suggestiveness of all this was plain. And now, at the far end of that cylindrical shed, John Endlich found the square, black-enamelled case, where all of those copper ribbons came together.

It was sealed, and apparently self-contained. Nothing could have damaged it very much, in the frigid stillness of millions of years. Its secrets were hidden within it. But they could not be too unfamiliar. And its presence was logical. A small, compact power unit. Nervously, he turned a little wheel. A faint vibration was transmitted to his gloved hand. And the globe in the ceiling began to glow.

He shut the thing off again. But how long did it take him to run back to his sagging creation of clear plastic, while the kids howled gleefully around him, and return with the end of a long cable, and pliers? How long did it take him to disconnect all of the glazed copper ribbons, and substitute the wires of the cable—attaching them to queer terminal-posts? No—not long.

The power was not as great as that which his own large atomic battery would have supplied. But it proved sufficient. And the current was direct—as it was supposed to be. The electrolysis apparatus bubbled vigorously. Slowly the tentlike roof began to rise, under the beginnings of a tiny gas-pressure.

"That does it, Pops!" Bubs shrilled.

"Yeah—maybe so," John Endlich agreed almost optimistically. He felt really tender toward his kids, just then. They'd really helped him, for once.

Yes—almost he was hopeful. Until he glanced at the rapidly declining sun. An all-night vigil. No. Probably worse. Oh Lord—how long could he last like this? Even if he managed to keep Neely and Company at bay? Night after night.... All that he had accomplished seemed useless. He just had so much more that could be wrecked—pushed over with a harsh laugh, as if it really was something funny.

John Endlich's flesh crawled. And in his thinking, now, he went a little against his own determinations. Probably because, in the present state of his disgust, he needed a drink—bad.

"Nuts!" he growled lugubriously. "If I'd only been a little more sociable.... That was where the trouble started. I might have got broke, but I would've made friends. They think I'm snooty."

Rose's jaw hardened, as if she took his regrets as an accusation that she had led him along the straight and narrow path, which—by an exasperating shift in philosophical principle—now seemed the shortest route to destruction. But he felt very sorry for her, too; and he didn't believe that what he had just said was entirely the truth.

So he added: "I don't mean it, Honey. I'm just griping."

She softened. "You've got to eat, Johnny," she said. "You haven't eaten all day. And tonight you've got to sleep. I'll keep watch. Maybe it'll be all right...."

Well, anyway it was nice to know that his wife was like that. Yeah—gentle, and fairminded. After they had all eaten supper, he tried hard to keep awake. Fear helped him to do so more than ever. Their tent was now covered by the rising plastic roof—but beyond the clear substance, he could still watch for starlight to be stopped by prowling forms, out there at the jagged rim of Vesta. It was hell to feel your skin puckering, and yet to have exhaustion pushing your eyelids down inexorably....

Somewhere he lost the hold on himself. And he dreamed that Alf Neely and he were fighting with their fists. And he was being beaten to a pulp. But he was wishing desperately that he could win. Then they could have a drink, and maybe be friends. But he knew hopelessly that things weren't quite that simple, either.

He awoke to blink at blazing sunshine. Then his whole body became clammy with perspiration, as he thought of his lapse from responsibility; glancing over, he saw that Rose was sleeping as soundly as the kids. His wide eyes searched for the disaster that he knew he'd find....

But the wide roof was all the way up, now — intact. It made a great, squarish bubble, the skin of which was specially treated to stop the hard and dangerous part of the ultra-violet rays of the sun, and also the lethal portion of the cosmic rays. It even had an inter-skin layer of gum that could seal the punctures that grain-of-sand-sized meteors might make. But meteors, though plentiful in the asteroid belt, were curiously innocuous. They all moved in much the same direction as the large asteroids, and at much the same velocity — so their relative speed had to be low.

The walls of the small tent around Endlich sagged, where they had bulged tautly before — showing that there was now a firm and equal pressure beyond them. The electrolysis apparatus had been left active all night, and the heating units. This was the result.

John Endlich was at first almost unbelieving when he saw that nothing had been wrecked during the night. For a moment he was elated. He woke up his family by shouting: "Look! The bums stayed away! They didn't come! Look! We've got five acres of ground, covered by air that we can breathe!"

His sense of triumph, however, was soon dampened. Yes — he'd been left unmolested — for one night. But had that been done only to keep him at a fruitless and sleepless watch? Probably. Another delicate form of hazing. And it meant nothing for the night to come — or for those to follow. So he was in the same harrowing position as before, pursued only by a wild and defenseless drive to get things done. To find some slight illusion of security by working to build a sham of normal, Earthly life. To shut out the cold vacuum, and a little of the bluntness of the voidal stars. To make certain reassuring sounds possible around him.

"Got to patch up the pieces of the house, first, and bolt 'em together, Rose," he said feverishly. "Kids — maybe you could help by setting out some of the hydroponic troughs for planting. We gotta break plain ground, too, as soon as it's thawed enough. We gotta...." His words raced on with his flying thoughts.

It was a mad day of toil. The hours were pitifully short. They couldn't be stretched to cover more than a fraction of all the work that Endlich wanted to get done. But the low gravity reduced the problem of heavy lifting to almost zero, at least. And he did get the house assembled — so that Rose and the kids and he could sleep inside its sealed doors. Sealed, that is, if Neely or somebody didn't use a blaster or an explosive cap or bullet—in an orgy of perverted humor.... He still had no answer for that.

Rose and the children toiled almost as hard as he did. Rose even managed to find a couple of dozen eggs, that—by being carefully packed to withstand a spaceship's takeoff—had withstood the effects of Neely's idea of fun. She set up an incubator, and put them inside, to be hatched.

But, of course, sunset came again—with the same pendent threat as before. Nerve-twisting. Terrible. And a vigil was all but impossible. John Endlich was out on his feet—far more than just dog-tired....

"That damned Neely," he groaned, almost too weary even to swallow his food, in spite of the luxury of a real, pullman-style supper table. "He doesn't lose sleep. He can pick his time to come here and raise hob!"

Rose's glance was strange—almost guilty. "Tonight I think he might have to stay home—too," she said.

John Endlich blinked at her.

"All right," she answered, rather defensively. "So to speak, Johnny, I called the cops. Yesterday—with the small radio transmitter. When you and Bubs and Evelyn were up in those old buildings. I reported Neely and his companions."

"Reported them?"

"Sure. To Mr. Mahoney, the boss at the mining camp. I was glad to find out that there is a little law and order around here. Mr. Mahoney was nice. He said that he wouldn't be surprised if they were cooled in the can for a few days, and then confined to the camp area. Matter of fact, I radioed him again last night. It's been done."

John Endlich's vast sigh of relief was slightly tainted by the idea that to call on a policing power for protection was a little bit on the timid side.

"Oh," he grunted. "Thanks. I never thought of doing that."

"Johnny."

"Yeah?"

"I kind of got the notion, though—from between the lines of what Mr. Mahoney said—that there was heavy trouble brewing at the camp. About conditions, and home-leaves, and increased profit-sharing. Maybe there's danger of riots and what-not, Johnny. Anyhow, Mr. Mahoney said that we should 'keep on exercising all reasonable caution.'"

"Hmm-m—Mr. Mahoney is *very* nice, ain't he?" Endlich growled.

"You stop that, Johnny," Rose ordered.

But her husband had already passed beyond thoughts of jealousy. He was thinking of the time when Neely would have worked out his sentence, and would be free to roam around again—no doubt with increased annoyance at the Endlich clan for causing his restraint. If a riot or something didn't spring him, beforehand. John Endlich itched to try to tear his head off. But, of course, the same consequences as before still applied....

As it turned out, the Endlichs had a reprieve of two months and fourteen days, almost to the hour and figured on a strictly Earth-time scale.

For what it was worth, they accomplished a great deal. In their great plastic greenhouse, supported like a colossal bubble by the humid, artificially-warmed air inside it, long troughs were filled with pebbles and hydroponic solution. And therein tomatoes were planted, and lettuce, radishes, corn, onions, melons—just about everything in the vegetable line.

There remained plenty of ground left over from the five acres, so John Endlich tinkered with that fifty-million-year-old tractor, figured out its atomic-power-to-steam principle, and used it to help harrow up the ancient soil of a smashed planet. He added commercial fertilizers and nitrates to it—the nitrates were, of course, distinct from the gaseous nitrogen that had been held, spongelike, by the subsoil, and had helped supply the greenhouse with atmosphere. Then he harrowed the ground again. The tractor worked fine, except that the feeble gravity made the lugs of its wheels slip a lot. He repeated his planting, in the old-fashioned manner.

Under ideal conditions, the inside of the great bubble was soon a mass of growing things. Rose had planted flowers—to be admired, and to help out the hive of bees, which were essential to some of the other plants, as well. Nor was the flora limited to the Earthly. Some seeds or spores had survived, here, from the mother world of the asteroids. They came out of their eons of suspended animation, to become root and tough, spiky stalk, and to mix themselves sparsely with vegetation that had immigrated from Earth, now that livable conditions had been restored over this little piece of ground. But whether they were fruit or weed, it was difficult to say.

Sometimes John Endlich was misled. Sometimes, listening to familiar sounds, and smelling familiar odors, toward the latter part of his reprieve, he almost imagined that he'd accomplished his basic desires here on Vesta—when he had always failed on Earth.

There was the smell of warm soil, flowers, greenery. He heard irrigation water trickling. The sweetcorn rustled in the wind of fans he'd set up to circulate the air. Bees buzzed. Chickens, approaching adolescence, peeped contentedly as they dusted themselves and stretched luxuriously in the shadows of the cornfield.

For John Endlich it was all like the echo of a somnolent summer of his boyhood. There was peace in it: it was like a yearning fulfilled. An end of wanderlust for him, here on Vesta. In contrast to the airless desolation outside, the interior of this five-acre greenhouse was the one most desirable place to be. So, except for the vaguest of stirrings sometimes in his mind, there was not much incentive to seek fun elsewhere. If he ever had time.

And there was a lot of the legendary, too, in what his family and he had accomplished. It was like returning a little of the blue sky and the sounds of life to this land of ruins and roadways and the ghosts of dead beauty. Maybe there'd be a lot more of all that, soon, when the rumored major influx of homesteaders reached Vesta.

"Yes, Johnny," Rose said once. "'Legendary' is a lot nicer word than 'ghostly'. And the ghosts are changing their name to legends."

Rose had to teach the kids their regular lessons. That children would be taught was part of the agreement you had to sign at the A. H. O. before you could be shipped out with them. But the kids had time for whimsy, too. In make-believe, they took their excursions far back to former ages. They played that they were "Old People."

Endlich, having repaired his atomic battery, didn't draw power anymore from the unit that had supplied the ancient buildings. But the relics remained. From a device like a phonograph, there was even a bell-like voice that chanted when a lever was pressed.

And it was the kids who found the first "tay-tay bug," a day after its trills were heard from among the new foliage. "Ta-a-a-ay-y-y — ta-a-a-a-ay-y-yy-y —" The sound was like that of a little wheel, humming with the speed of rotation, and then slowing to a scratchy stop.

A one-legged hopper, with a thin but rigid gliding wing of horn. Opalescent in its colors. It had evidently hatched from a tiny egg, preserved by the cold for ages.

Wise enough not to clutch it with his bare hands, Bubs came running with it held in a leaf.

It proved harmless. It was ugly and beautiful. Its great charm was that it was a vocal echo from the far past.

Sure. Life got to be fairly okay, in spite of hard work. The Endlichs had conquered the awful stillness with life-sounds. Growing plants kept the air in their greenhouse fresh and breathable by photosynthesis. John Endlich did a lot of grinning and whistling. His temper never flared once. Deep down in him there was only a brooding certainty that the calm couldn't last. For, from all reports, trouble seethed at the mining camp. At any time there might be a blowup, a reign of terror that would roll over all of Vesta. A thing to release pent-up forces in men who had seen too many hard stars, and had heard too much stillness. They were like the stuff inside a complaining volcano.

The Endlichs had sought to time their various crops, so that they would all be ready for market on as nearly as possible the same day. It was intended as a trick of advertising—a dramatically sudden appearance of much fresh produce.

So, one morning, in a jet-equipped space-suit, Endlich arced out for the mining camp. Inside the suit he carried samples from his garden. Six tomatoes. Beauties.

"Have luck with them, Johnny! But watch out!" Rose flung after him by helmet phone. With a warm laugh. Just for a moment he felt maybe a little silly. Tomatoes! But they were what he was banking on, and had forced toward maturity, most. The way he figured, they were the kind of fruit that the guys in the camp—gagged by a diet of canned and dehydrated stuff, because they were too busy chasing mineral wealth to keep a decent hydroponic garden going—would be hungriest for.

Well—he was rather too right, in some ways, to be fortunate. Yeah—they still call what happened the Tomato War.

Poor Johnny Endlich. He was headed for the commissary dome to display his wares. But vague urges sidetracked him, and he went into the recreation dome of the camp, instead.

And into the bar.

The petty sin of two drinks hardly merits the punishing trouble which came his way as, at least partially, a result. With his face-window open, he stood at the bar with men whom he had never seen before. And he began to have minor delusions of grandeur. He became a little too proud of his accomplishments. His wariness slipped into abeyance. He had a queer idea that, as a farmer with concrete evidence of his skills to show, he would win respect that had been denied him. Dread of consequences of some things that he might do, became blurred. His hot temper began to smolder, under the spark of memory and the fury of insult and malicious tricks, that, considering the safety of his loved ones, he had had no way to fight back against. Frustration is a dangerous force. Released a little, it excited him more. And the tense mood of the camp—a thing in the very air of the domes—stirred him up more. The camp—ready to explode into sudden, open barbarism for days—was now at a point where nothing so dramatic as fresh tomatoes and farmers in a bar was needed to set the fireworks off.

John Endlich had his two drinks. Then, with calm and foolhardy detachment, he set the six tomatoes out in a row before him on the synthetic mahogany.

He didn't have to wait at all for results. Bloodshot eyes, some of them belonging to men who had been as gentle as lambs in their ordinary lives on Earth, turned swiftly alert. Bristly faces showed swift changes of expression: surprise, interest, greed for possession—but most of all, aggressive and Satanic humor.

"*Jeez — tamadas!*" somebody growled, amazed.

Under the circumstances, to be aware of opportunity was to act. Big paws, some bare and calloused, some in the gloves of space suits, reached out, grabbed. Teeth bit. Juice squirted, landing on hard metal shaped for the interplanetary regions.

So far, fine. John Endlich felt prouder of himself—he'd expected a certain fierceness and lack of manners. But knowing all he did know, he should have taken time to visualize the inevitable chain-reaction.

"Thanks, pal.... You're a prince...."

Sure—but the thanks were more of a mockery than a formality.

"Hey! None for me? Whatsa idea?..."

"Shuddup, Mic.... Who's dis guy?... Say, Friend—you wouldn't be that pun'kin-head we been hearin' about, would you?... Well—my gracious— bet you are! Dis'll be nice to watch!..."

"Where's Alf Neely, Cranston? What we need is excitement."

"Seen him out by the slot-machines. The bar is still out of bounds for him. He can't come in here."

"Says who? Boss Man Mahoney? For dis much sport Neely can go straight to hell! And take Boss Man with him on a pitchfork.... Hey-y-y!... Ne-e-e-e-l-y-y-y!..."

The big man whose name was called lumbered to the window at the entrance to the bar, and peered inside. During the last couple of months he'd been in a perpetual grouch over his deprivation of liberty, which had rankled him more as an affront to his dignity.

When he saw the husband of the authoress of his woes — the little bum, who, being unable to guard his own, had allowed his woman to holler "Cop!" — Neely let out a yell of sheer glee. His huge shoulders hunched, his pendulous nose wobbled, his squinty eyes gleamed and he charged into the bar.

John Endlich's first reaction was curiously similar to Neely's. He felt a flash of savage triumph under the stimulus of the thought of immediate battle with the cause of most of his troubles. Temper blazed in him.

Belatedly, however, the awareness came into his mind that he had started an emotional avalanche that went far beyond the weight and fury of one man like Neely. Lord, wouldn't he ever learn? It was tough as hell to crawl, but how could a man put his wife and kids in awful jeopardy at the hands of a flock of guys whom space had turned into gorillas?

Endlich tried for peace. It was to his credit that he did so quite coolly. He turned toward his charging adversary and grinned.

"Hi, Neely," he said. "Have a drink — on me."

The big man stopped short, almost in unbelief that anyone could stoop so low as to offer appeasement. Then he laughed uproariously.

"Why, I'd be delighted, Mr. Pun'kins," he said in a poisonous-sweet tone. "Let bygones be bygones. Hey, Charlie! Hear what Pun'kins says? The drinks are all on him! And how is the Little Lady, Mrs. Pun'kins? Lonesome, I bet. Glad to hear it. I'm gonna fix that!"

With a sudden lunge Neely gripped Endlich's hand, and gave it a savage if momentary twist that sent needles of pain shooting up the homesteader's arm. It was a goading invitation to battle, which grim knowledge of the sequel now compelled Endlich to pass up.

"Don't call him Pun'kins, Neely!" somebody yelled. "It ain't polite to mispronounce a name. It's Mr. Tomatoes. I just saw. Bet he's got a million of 'em, out there on the farm!"

The whole crowd in the bar broke into coarse shouts and laughs and comments. "... We ain't good neighbors — neglecting our social duties. Let's pay 'em a visit.... Pun'kins! What else you got besides tamadas? Let's go on a picnic!... Hell with the Boss Man!... Yah-h-h — We need some diversion.... I'm not goin' on shift.... Come on, everybody! There's gonna be a fight — a moider!... Hell with the Boss Man...."

Like the flicker of flame flashing through dry gunpowder, you could feel the excitement spread. Out of the bar. Out of the rec-dome. It would soon ignite the whole tense camp.

John Endlich's heart was in his mouth, as his mind pictured the part of all this that would affect him and his. A bunch of men gone wild, kicking over the traces, arcing around Vesta, sacking and destroying in sheer exuberance, like brats on Hallowe'en. They would stop at nothing. And Rose and the kids....

This was it. What he'd been so scared of all along. It was at least partly his own fault. And there was no way to stop it now.

"I love tomatoes, Mr. Pun'kins," Neely rumbled at Endlich's side, reaching for the drink that had been set before him. "But first I'm gonna smear you all over the camp.... Take my time—do a good job.... Because y'didn't give me any tomatoes...."

Whereat, John Endlich took the only slender advantage at hand for him—surprise. With all the strength of his muscular body, backed up by dread and pent-up fury, he sent a gloved fist crashing straight into Neely's open face-window. Even the pang in his well-protected knuckles was a satisfaction—for he knew that the damage to Neely's ugly features must be many times greater.

The blow, occurring under the conditions of Vesta's tiny gravity, had an entirely un-Earthly effect. Neely, eyes glazing, floated gently up and away. And Endlich, since he had at the last instant clutched Neely's arm, was drawn along with the miner in a graceful, arcing flight through the smoky air of the bar. Both armored bodies, lacking nothing in inertia, tore through the tough plastic window, and they bounced lightly on the pavement of the main section of the rec-dome.

Neely was as limp as a wet rag, sleeping peacefully, blood all over his crushed face. But that he was out of action signified no peace, when so many of his buddies were nearby, and beginning to seethe, like a swarm of hornets.

So there was an element of despair in Endlich's quick actions as he slammed Neely's face-window and his own shut, picked up his enemy, and used his jets to propel him in the long leap to the airlock of the dome. He had no real plan. He just had the ragged and all but hopeless thought of using Neely as a hostage—as a weapon in the bitter and desperate attempt to defend his wife and children from the mob that would be following close behind him....

Tumbling end over end with his light but bulky burden, he sprawled at the threshold of the airlock, where the guard, posted there, had stepped hastily out of his way. Again, capricious luck, surprise, and swift action were on his side. He pressed the control-button of the lock, and squirmed through its double valves before the startled guard could stop him.

Then he slammed his jets wide, and aimed for the horizon.

It was a wild journey—for, to fly straight in a frictionless vacuum, any missile must be very well balanced; and the inertia and the slight but unwieldy weight of Neely's bulk disturbed such balance in his own jet-equipped space suit. The journey was made, then, not in a smooth arc, but in a series of erratic waverings. But what Endlich lacked in precise direction, he made up in sheer reckless, dread-driven speed.

From the very start of that wild flight, he heard voices in his helmet phones:

"Damn pun'kin-head greenhorn! Did you see how he hit Neely, Schmidt? Yeah—by surprise.... Yeah—Kuzak. I saw. He hit without warning.... Damn yella yokel.... Who's comin' along to get him?..."

Sure—there was another side to it—other voices:

"Shucks—Neely had it coming to him. I hope the farmer really murders that big lunkhead.... You ain't kiddin', Muir. I was glad to see his face splatter like a rotten tamata...."

Okay—fine. It was good to know you had some sensible guys on your side. But what good was it, when the camp as a whole was boiling over from its internal troubles? There were more than enough roughnecks to do a mighty messy job—fast.

Panting with tension, Endlich swooped down before his greenhouse, and dragged Neely inside through the airlock. For a fleeting instant the sights and sounds and smells that impinged on his senses, as he opened his face-window once more, brought him a regret. The rustle of corn, the odor of greenery, the chicken voices—there was home in all of this. Something pastoral and beautiful and orderly—gained with hard work. And something brought back—restored—from the remote past. The buzzing of the tay-tay bug was even a real echo from that smashed yet undoubtedly once beautiful world of antiquity.

But these were fragile concerns, beside the desperate question of the immediate safety of Rose and the kids.... Already cries and shouts and comments were coming faintly through his helmet phones again:

"Get the yokel! Get the bum!... We'll fix his wagon good...."

The pack was on the way—getting closer with every heartbeat. Never in his life had Endlich experienced so harrowing a time as this; never, if by some miracle he lived, could he expect another equal to it.

To stand and fight, as he would have done if he were alone, would mean simply that he would be cut down. To try the peacemaking of appeasement, would have probably the same result—plus, for himself, the dishonor of contempt.

So, where was there to turn, with grim, unanswering blankness on every side?

John Endlich felt mightily an old yearning—that of a fundamentally peaceful man for a way to oppose and win against brutal, overpowering odds without using either serious violence or the even more futile course of supine submission. Here on Vesta, this had been the issue he had faced all along. In many ages and many nations — and probably on many planets throughout the universe — others had faced it before him.

To his straining and tortured mind the trite and somewhat mocking answers came: Psychology. Salesmanship. The selling of respect for one's self.

Ah, yes. These were fine words. Glib words. But the question, "How?" was more bitter and derisive than ever.

Still, he had to try something — to make at least a forlorn effort. And now, from certain beliefs that he had, coupled with some vague observations that he had made during the last hour, a tattered suggestion of what form that effort might take, came to him.

As for his personal defects that had given him trouble in the past — well — he was lugubriously sure that he had learned a final lesson about liquor. For him it always meant trouble. As for wanderlust, and the gambling and hell-raising urge — he had been willing to stay put on Vesta, named for the goddess of home, for weeks, now. And he was now about to make his last great gamble. If he lost, he wouldn't be alive to gamble again. If, by great good-fortune, he won — well he was certain that all the charm of unnecessary chance-taking would, by the memory of these awful moments, be forever poisoned in him.

Now Rose and the youngsters came hurrying toward him.

"Back so soon, Johnny?" Rose called. "What's this? What happened?"

"Who's the guy, Pop?" Evelyn asked. "Oh — Baloney Nose.... What are you doing with him?"

But by then they all had guessed some of the tense mood, and its probable meaning.

"Neely's pals are coming, Honey," Endlich said quietly. "It's the showdown. Hide the kids. And yourself. Quick. Under the house, maybe."

Rose's pale eyes met his. They were comprehending, they were worried, but they were cool. He could see that she didn't want to leave him.

Evelyn looked as though she might begin to whimper; but her small jaw hardened.

Bubs' lower lip trembled. But he said valiantly: "I'll get the guns, Pop, I'm stayin' with yuh."

"No you're not, son," John Endlich answered. "Get going. Orders. Get the guns to keep with you — to watch out for Mom and Sis."

Rose took the kids away with her, without a word. Endlich wondered how to describe what was maybe her last look at him. There were no fancy words in his mind. Just Love. And deep concern.

Alf Neely was showing signs of returning consciousness. Which was good. Still dragging him, Endlich went and got a bushel basket. It was filled to the brim with ripe, red tomatoes, but he could carry its tiny weight on the palm of one hand, scarcely noticing that it was there.

For an instant Endlich scanned the sky, through the clear plastic roof of the great bubble. He saw at least a score of shapes in space armor, arcing nearer — specks in human form, glowing with reflected sunlight, like little hurtling moons among the stars. Neely's pals. In a moment they would arrive.

Endlich took Neely and the loaded basket close to the transparent side of the greenhouse, nearest the approaching roughnecks. There he removed Neely's oxygen helmet, hoping that, maybe, this might deter his friends a little from rupturing the plastic of the huge bubble and letting the air out. It was a feeble safeguard, for, in all probability, in case of such rupture, Neely would be rescued from death by smothering and cold and the boiling of his blood, simply by having his helmet slammed back on again.

Next, Endlich dumped the contents of the basket on the ground, inverted it, and sat Neely upon it. The big man had recovered consciousness enough to be merely groggy by now. Endlich slapped his battered face vigorously, to help clear his head — after having, of course, relieved him of the blaster at his belt.

Endlich left his own face-window open, so that the sounds of Neely's voice could penetrate to the mike of his own helmet phone, thus to be transmitted to the helmet phones of Neely's buddies.

Endlich was anything but calm inside, with the wild horde, as irresponsible in their present state of mind as a pack of idiot baboons, bearing down on him. But he forced his tone to be conversational when he spoke.

"Hello, Neely," he said. "You mentioned you liked tomatoes. Maybe you were kidding. Anyhow I brought you along home with me, so you could have some. Here on the ground, right in front of you, is a whole bushel. The regular asteroids price — considering the trouble it takes to grow 'em, and the amount of dough a guy like you can make for himself out here, is five bucks apiece. But for you, right now, they're all free. Here, have a nice fresh, ripe one, Neely."

The big man glared at his captor for a second, after he had looked dazedly around. He would have leaped to his feet — except that the muzzle of his own blaster was leveled at the center of his chest, at a range of not over twenty inches. For a fleeting instant, Neely looked scared and prudent. Then he saw his pals, landing like a flock of birds, just beyond the transparent side of the greenhouse. And he heard their shouts, coming loudly from Endlich's helmet-phones:

"We come after you, Neely! We'll get the damn yokel off your neck.... Come on, guys—let's turn the damn place upside down!..."

Neely grew courageous—yes, maybe it did take a certain animal nerve to do what he did. His battered and bloodied lip curled.

"Whatdayuh think you're up to, Pun'kin-head!" he snarled slowly, his tone dripping contempt for the insanely foolish. He laughed sourly, "Haw-haw-haw." Then his face twisted into a confident and mocking leer. To carry the mockery farther, a big paw reached out and grabbed the proffered tomato from Endlich's hand. "Sure—thanks. Anything to oblige!" He took a great bite from the fruit, clowning the action with a forced expression of relish. "Ummm!" he grunted. In danger, he was being the showman, playing for the approval of his pals. He was proving his comic coolness—that even now he was master of the situation, and was in no hurry to be rescued. "Come on, punk!" he ordered Endlich. "Where is the next one, seeing you're so generous? Be polite to your guest!"

Endlich handed him a second tomato. But as he did so, it seemed all the things he dreaded would happen were breathing down his back. For the faces that he glimpsed beyond the plastic showed the twisted expressions that betray the point where savage humor imperceptibly becomes murderous. A dozen blasters were leveled at him.

But the eyes of the men outside showed, too, the kind of interest that any odd procedure can command. They stood still for a moment, watching, commenting:

"Hey—Neely! See if you can down the next one with one bite!... Don't eat 'em all, Neely! Save some for us!..."

Endlich was following no complete plan. He had only the feeling that somewhere here there might be a dramatic touch that, by a long chance, would yield him a toehold on the situation. Without a word, he gave Neely a third tomato. Then a fourth and a fifth....

Neely kept gobbling and clowning.

Yeah—but can this sort of horseplay go on until one man has consumed an entire bushel of tomatoes? The question began to shine speculatively in the faces of the onlookers. It began to appeal to their wolfish sense of comedy. And it started to betray itself—in another manner—in Neely's face.

After the fifteenth tomato, he burped and balked. "That's enough kiddin' around, Pun'kin-head," he growled. "Get away with your damned garden truck! I should be beatin' you to a grease-spot right this minute! Why — I — "

Then Neely tried to lunge for the blaster. As Endlich squeezed the trigger, he turned the weapon aside a trifle, so that the beam of energy flicked past Neely's ear and splashed garden soil that turned incandescent, instantly.

John Endlich might have died in that moment, cut down from behind. That he wasn't probably meant that, from the position of complete underdog among the spectators, his popularity had risen some.

"Neely," he said with a grin, "how can you start beatin', when you ain't done eatin'? Neely — here I am, trying to be friendly and hospitable, and you aren't co-operating. A whole bushel of juicy tomatoes — symbols of civilization way the hell out here in the asteroids — and you haven't even made a dent in 'em yet! What's the matter, Neely? Lose your appetite? Here! Eat!..."

Endlich's tone was falsely persuasive. For there was a steely note of command in it. And the blaster in Endlich's hand was pointed straight at Neely's chest.

Neely's eyes began to look frightened and sullen. He shifted uncomfortably, and the bushel basket creaked under his weight. "You're yella as any damn pun'kin!" he said loudly. "You don't fight fair!... Guys — what's the matter with you? Get this nut with the blaster offa me!..."

"Hmm — yella," Endlich seemed to muse. "Maybe not as yella as you were once — coming around here at night with a whole gang, not so long ago — "

"Call *me* yella?" Nelly hollered. "Why, you lousy damn yokel, if you didn't have that blaster — "

Endlich said grimly, "But I got it, friend!" He sent a stream of energy from the blaster right past Neely's head, so close that a shock of the other's hair smoked and curled into black wisps. "And watch your language — my wife and kids can hear you — "

Neely's thick shoulders hunched. He ducked nervously, rubbing his head — and for the first time there was a hint of genuine alarm in his voice. "All right," he growled, "all right! Take it easy — "

Something deep within John Endlich relaxed — a cold tight knot seemed to unwind — for, at that moment, he knew that Neely was beginning to lose. The big man's evident discomfort and fear were the marks of weakness — to his followers at least; and with them, he could never be a leader, again. Moreover, he had allowed himself to be maneuvered into the position of being the butt of a practical joke, that, by his own code, must be followed up, to its nasty, if interesting, outcome. The spectators began to resemble Romans at the circus, with Neely the victim. And the victim's downfall was tragically swift.

"Come on, Neely! You heard what Pun'kins said," somebody yelled. "Jeez—a whole bushel. Let's see how many you can eat, Neely.... Damned if this ain't gonna be rich! Don't let us down, Neely! Nobody's hurtin' yuh. All you have to do is eat—all them nice tamadas.... Hey, Neely—if that bushel ain't enough for you, I'll personally buy you another, at the reg'lar price. Haw-haw-haw.... Lucky Neely! Look at him! Having a swell banquet. Better than if he was home.... Haw-haw-haw.... Come on, Pun'kins—make him eat!..."

Yeah, under certain conditions human nature can be pretty fickle. Wonderingly, John Endlich felt himself to be respected—the Top Man. The guy who had shown courage and ingenuity, and was winning, by the harsh code of men who had been roughened and soured by space—by life among the asteroids.

For a little while then, he had to be hard. He thrust another tomato toward Neely, at the same time directing a thin stream from the blaster just past the big nose. Neely ate six more tomatoes with a will, his eyes popping, sweat streaming down his forehead.

Endlich's next blaster-stream barely missed Neely's booted toe. The persuasive shot was worth fifty-five more dollars in garden fruit consumed. The crowd gave with mock cheers and bravos, and demanded more action.

"That makes thirty-two.... Come on, Neely—that's just a good start. You got a long, long ways to go.... Come on, Pun'kins—bet you can stuff fifty into him...."

To goad Neely on in this ludicrous and savage game, Endlich next just scorched the metal at Neely's shoulder. It isn't to be said that Endlich didn't enjoy his revenge—for all the anguish and real danger that Neely had caused him. But as this fierce yet childish sport went on, and the going turned really rough for the big asteroid miner, Endlich's anger began to be mixed with self-disgust. He'd always be a hot-tempered guy; he couldn't help that. But now, satisfaction, and a hopeful glimpse of peace ahead, burned the fury out of him and touched him with shame. Still, for a little more, he had to go on. Again and again, as before, he used that blaster. But, as he did so, he talked, ramblingly, knowing that the audience, too, would hear what he said. Maybe, in a way, it was a lecture; but he couldn't help that:

"Have another tomato, Neely. Sorry to do things like this—but it's your own way. So why should you complain? Funny, ain't it? A man can get even too many tomatoes. Civilized tomatoes. Part of something most guys around here have been homesick for, for a long time.... Maybe that's what has been most of the trouble out here in the asteroids. Not enough civilization. On Earth we were used to certain standards—in spite of being rough enough there, too. Here, the traces got kicked over. But on this side of Vesta, an idea begins to soak in: This used to be nice country—blue sky, trees growing. Some of that is coming back, Neely. And order with it. Because, deep in our guts, that's what we all want. And fresh vegetables'll help.... Have another tomato, Neely. Or should we call it enough, guys?"

"Neely, you ain't gonna quit now?" somebody guffawed. "You're doin' almost good. Haw-haw!"

Neely's face was purple. His eyes were bloodshot. His mouth hung partly open. "Gawd—no—please!" he croaked.

An embarrassed hush fell over the crowd. Back home on Earth, they had all been more-or-less average men. Finally someone said, expressing the intrusion among them of the better dignity of man:

"Aw—let the poor dope go...."

Then and there, John Endlich sold what was left of his first bushel of tomatoes. One of his customers—the once loud-mouthed Schmidt—even said, rather stiffly, "Pun'kins—you're all right."

And these guys were the real roughnecks of the mining camp.

Is it necessary to mention that, as they were leaving, Neely lost his pride completely, soiling the inside of his helmet's face-window so that he could scarcely see out of it? That, amid the raucous laughter of his companions, which still sounded slightly self-conscious and pitying. Thus Alf Neely sank at last to the level of helpless oblivion and nonentity.

A week of Vestal days later, in the afternoon, Rose and the kids came to John Endlich, who was toiling over his cucumbers.

"Their name is Harper, Pop!" Bubs shouted.

"And they've got three children!" Evelyn added.

John Endlich, straightened, shaking a kink out of his tired back. "Who?" he questioned.

"The people who are going to be our new neighbors, Johnny," Rose said happily. "We just picked up the news on the radio—from their ship, which is approaching from space right now! I hope they're nice folks. And, Johnny — there used to be country schools with no more than five pupils...."

"Sure," John Endlich said.

Something felt warm around his heart. Leave it to a woman to think of a school — the symbol of civilization, marching now across the void. John Endlich thought of the trouble at the mining camp, which his first load of fresh vegetables, picked up by a small space boat, had perhaps helped to end. He thought of the relics in this strange land. Things that were like legends of a lost pastoral beauty. Things that could come back. The second family of homesteaders was almost here. Endlich was reconciled to domesticity. He felt at home; he felt proud.

Bees buzzed near him. A tay-tay bug from a perished era, hummed and scraped out a mournful sound.

"I wonder if the Harper kids'll call you Mr. Pun'kins, Pop," Bubs remarked. "Like the miners still do."

John Endlich laughed. But somehow he was prouder than ever. Maybe the name would be a legend, too.

www.ingramcontent.com/pod-product-compliance
Lightning Source LLC
LaVergne TN
LVHW041806190726
843493LV00008B/2814